SCIENCE SERIES

Number

Geothermal Exploration of Western Trans-Pecos Texas

by

JERRY M. HOFFER

Department of Geological Sciences, The University of Texas at El Paso

TEXAS WESTERN PRESS
THE UNIVERSITY OF TEXAS AT EL PASO
1979

Library of Congress Catalog Card No. 78-57754

ISBN 87404-060-4

$3.00 per Copy

CONTENTS

Introduction 1

Location and Collection of Samples 4

Analytical Procedures 4

Thermal Waters in Trans-Pecos Texas 5

Thermal Water Occurrences 7

El Paso County 7

Hudspeth County 8

Culberson-Jeff Davis Counties 8

Presidio County 9

Brewster County 9

Temperatures of Thermal Occurrences 10

Area of Geothermal Potential 10

Northeastern El Paso County 13

Surface Temperatures 13

Silica Geothermal Temperatures 13

Thermal Gradient 15

Western Presidio County 15

Surface Temperatures 15

Silica Geothermal Temperatures 15

Marfa Area 23

Surface Temperatures 23

Silica Geothermal Temperatures 23

Jeff Davis and Brewster Counties 23

Chemistry of Thermal Waters 23

Geochemistry of Nonthermal Groundwaters 26
Chemistry of Water 26
Presentation of Analyses 26
Temperature and pH 27
Total Dissolved Solids 27
Major Constituents 29
Silicon 29
Calcium 31
Magnesium 33
Sodium 35
Chloride 38
Secondary Constituents 40
Fluoride 40
Potassium 40
Iron 41
Strontium 42
Minor Constituents 43
Barium 43
Lithium 43
Bromide and Iodide 44
Copper, Lead, and Zinc 44
Manganese 45
Cadmium 46
Trace Constituents 47
Silver 47
Future Studies 47
Acknowledgements 48
References 49

Geothermal Exploration of Western Trans-Pecos, Texas

GEOTHERMAL EXPLORATION OF WESTERN TRANS-PECOS, TEXAS

By

JERRY M. HOFFER

Geothermal refers to the natural heat of the earth. The earth is a great reservoir of heat, with all but the outer crust containing high temperatures. This heat of the earth is one of the most immense energy resources available to man.

Energy is the driving force of industry and of civilization. The sources of energy in the United States today include oil, natural gas, hydroelectric, coal, and minor nuclear power; the major sources are oil and natural gas, which are being depleted at an alarming rate. Energy sources for the United States in 1971 are shown in TABLE 1 and forecasts indicate that by 1985, oil and natural gas will provide over three-fourths of our energy, with most of the oil and gas being imported.

When recoverable, the heat of the earth can be employed to supply energy for man and thereby conserve our dwindling supply of fossil fuels. In several areas of the world, nature delivers geothermal energy to the earth's surface in the form of steam which can be used to generate electricity. The first commercial use of natural steam was at Larderello, Italy, where it has been used to generate electricity since 1904. The largest geothermal source of electricity is at the Geysers in California, in operation since 1960. Other geothermal power plants are in Mexico, Japan, New Zealand, Iceland, and the U.S.S.R.

In other areas of the world, waters from thermal springs or wells are hot enough to heat homes and greenhouses, supply energy for low temperature processes, desalinate water, and aid in mineral extraction.

TABLE 1

Energy Sources for the U.S. Economy in Trillions of BTU

ENERGY SOURCE	1971	FORECASTED 1985
Coal	12.6	21.5
Oil	30.5	50.7
Natural Gas	22.7	28.4
Nuclear Power	0.4	11.7
Hydroelectric	2.8	4.3
	69.0	116.6

Source: Otte and Kruger, 1973.

The economy of the area surrounding the Rio Grande Valley in west Texas is dependent upon energy and water. In this area, the availability of sufficient, good quality water is largely dependent on the availability of reasonably priced energy. Natural gas is the most important source of energy in this region; it is used for generating electricity, heating buildings, and for fueling pumps for irrigated agriculture.

This report summarizes the results of a geothermal research project undertaken from September 1, 1975 to August 31, 1979 by the Department of Geological Science, The University of Texas at El Paso. The area of study includes a six-county area of west Texas that encompasses over 22,000 square miles in El Paso, Hudspeth, Culberson, Jeff Davis, Presidio, and Brewster Counties (FIGURE 1). The purpose of the research effort was essentially twofold: 1) to locate and define the areas of highest geothermal potential in Trans-Pecos, Texas based on geochemical indicators, and

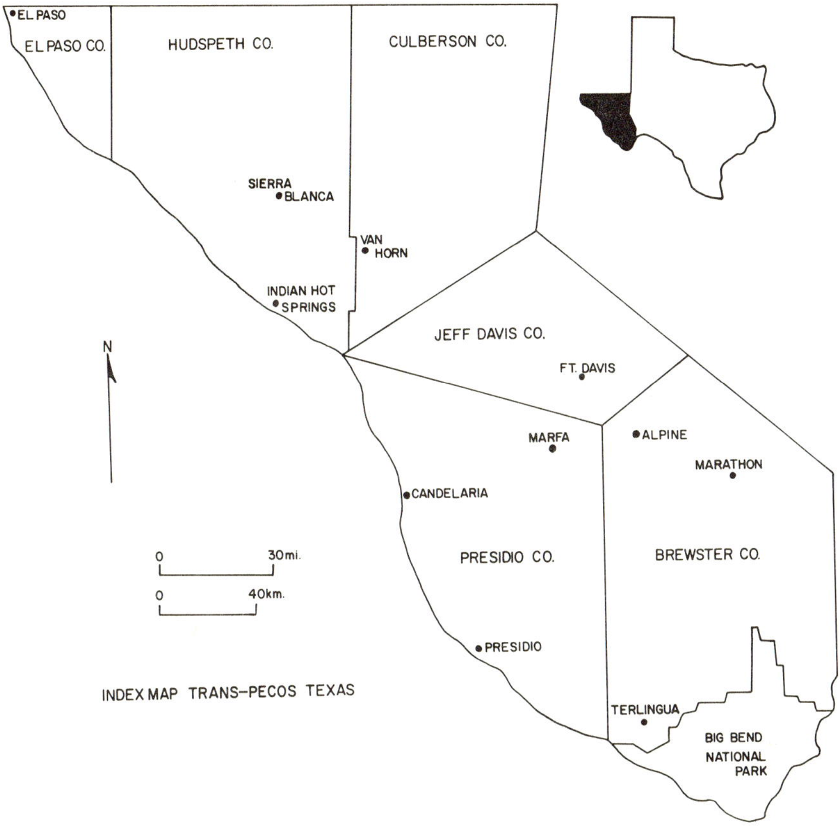

FIGURE 1. Index Map of the Trans-Pecos, Texas Area.

2) to characterize the chemistry of subsurface waters in the area for future mineral extraction and desalination projects.

Location and Collection of Samples

Water samples have been collected from all available springs, wells, and windmills in the Trans-Pecos region. As of September 1978, water samples totaling 1028 had been collected and analyzed. The distribution of samples in the six-county area is shown in TABLE 2.

At each site of collection, the temperature and pH of the water were measured and, where possible, the depth of the well was recorded. Samples were acidified in the field and were returned as soon as possible to the analytical laboratories in the Department of Geological Sciences, U. T. El Paso.

TABLE 2

Distribution of Collected Groundwater Samples in Trans-Pecos, Texas

SAMPLES	COUNTY
116	El Paso
166	Culberson
124	Hudspeth
161	Jeff Davis
225	Presidio
236	Brewster
1028	

Analytical Procedures

In the laboratory, all samples were analyzed for total dissolved solids, and the 19 individual chemical constituents are shown in TABLE 3. Equipment that was used for the analyses included an atomic-absorption spectrophotometer and specific-ion electrodes.

The analyses of potassium-sodium-calcium and silicon were used to calculate geothermometers. These geothermometers, as outlined by Fournier and Rowe (1966), are used to estimate the temperature of groundwater at any depth and are, therefore, used to locate subsurface hot water deposits.

TABLE 3

Chemical Ions Analyzed in Trans-Pecos, Texas Waters

A. Major Constituents (range of concentration 1 to 1000 ppm)

*Calcium	*Chloride	*Magnesium	*Silicon
*Sodium			

B. Secondary Constituents (range of concentration 0.01 to 10 ppm)

*Fluoride	*Iron	*Potassium	*Strontium

C. Minor Constituents (range of concentration 0.001 to 0.1 ppm)

*Barium	°Bromide	*Cadmium	*Copper
°Iodide	*Lead	*Lithium	*Manganese
*Zinc			

D. Trace Constituents (range of concentration less than 0.001 ppm)

*Silver

* - by atomic absorption spectrophotometer

° - by specific ion electrodes

Thermal Waters in Trans-Pecos, Texas

The definition of *thermal water* is arbitrary. According to Summers (1976), subsurface temperatures are affected by climatic conditions to a depth of about 100 feet; below this depth, temperatures increase an average of 1°C (1.8°F) every 100 feet; therefore, shallow-water temperatures are influenced by surface temperatures, and deep-water sources by the increase in temperature below the surface of the earth.

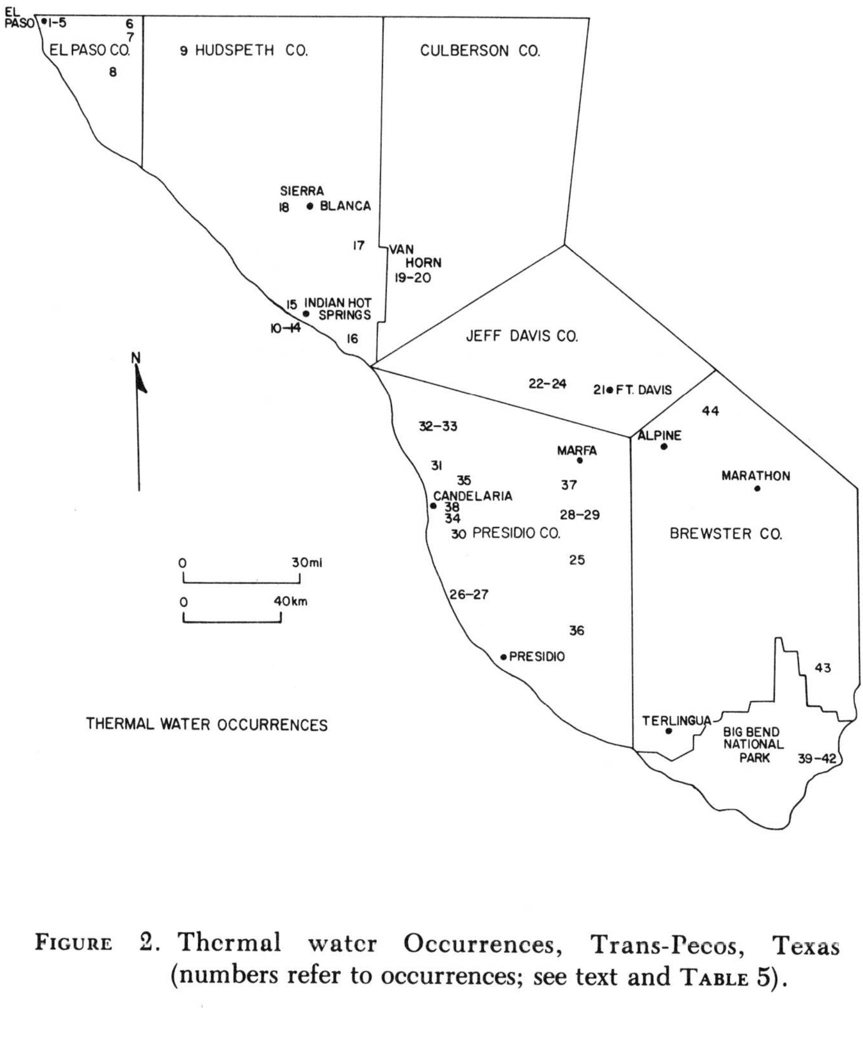

FIGURE 2. Thermal water Occurrences, Trans-Pecos, Texas (numbers refer to occurrences; see text and TABLE 5).

For this report, thermal waters are defined as those that have temperatures equal to or exceeding 30°C (86°F). This value is some 7°C (13°F) above the average subsurface-water temperature in west Texas, and approximately 11°C (20°F) above the mean annual air temperature of the region.

A number of warm-to-hot springs and wells have been reported along both sides of the Rio Grande in west Texas and northern Mexico (Groat, 1972; Warning, 1965). However, numerous unreported occurrences of thermal waters exist in the region, but little data are available concerning the occurrence and chemistry of these waters.

During this study, 44 occurrences of thermal waters were found in Trans-Pecos, Texas; these occurrences are shown in FIGURE 2. Temperatures of the waters range from 30°C (86°F) to 72°C (160°F); the hottest waters are found in northwest Presidio County at the Gulf wells. The average thermal-water temperature is approximately 37°C (99°F); distribution of thermal-water temperatures for each county is shown in TABLE 3.

Thermal Water Occurrences

El Paso County

Two areas in the county contain thermal waters (FIGURE 2). One area is near El Paso, just west of the Franklin Mountains, where the El Paso Water Utilities encountered warm water in six wells drilled in the Mesilla bolson. Only four wells were available for sampling during this study (locations 1-4, FIGURE 2). All warm-water wells produce from fine-grained sands of the Santa Fe Group, at temperatures averaging approximately 33°C (91°F). Average depth of the sampled wells is just over 1000 feet.

Two warm-to-hot water wells are located in northeastern El Paso County (locations 6 and 7, FIGURE 2). At location 6, water at 56°C (136°F) is encountered at a depth of 595 feet, in limestone probably of Permian age (Hueco Group). A few miles to the south at location 7, water at

40°C (104°F) is pumped from limestone at a depth of approximately 1000 feet. At location 8, water of 33°C (91°F) flows from a 400 foot well in bolson-fill sediments.

Hudspeth County

Warm-to-hot waters occur at Indian Hot Springs (locations 10 - 14, FIGURE 2), Red Bull Spring (location 15, FIGURE 2), Hot Wells (location 16, FIGURE 2), a city well at Sierra Blanca, (location 18, FIGURE 2) and a single well in the northwest part of the county (location 9, FIGURE 2).

Eight springs occur at Indian Hot Springs, all are located on the Rio Grande floodplain and seem to rise along the Caballo fault. The springs are located on the downthrown block, a short distance from the fault trace. The waters rise along the Caballo fault and discharge at the surface in shallow dug wells (Reaser, Underwood, and Jones, 1975). Five of them discharge warm water.

To the north, warm water from Red Bull Spring flows to the surface at the intersection of the Caballo and Red Bull faults (Jones, 1968). Hot Wells, located southeast of Sierra Blanca, are drilled into bolson-fill sediments. The two wells were drilled by the Southern Pacific Railroad to a depth of approximately 1000 feet. Only one well is currently in operation, and it was the only one sampled during this study.

The temperatures of thermal waters in southern Hudspeth County range from 32°C (90°F) to 46°C (115°F), with the hottest waters occurring at Chief Spring at Indian Hot Springs.

Culberson - Jeff Davis Counties

Two warm water wells of 32°C and 30°C are located in Culberson County a few miles southeast of Van Horn; the wells are 750 and 200 feet deep, respectively, (locations 19 and 20, FIGURE 2). Three warm springs and one warm well occur west of Fort Davis in Jeff Davis County; the warm springs average approximately 31°C (88°F), (locations 16-19, FIGURE 2). The springs seem to flow from faults in Tertiary volcanic units.

Presidio County

The largest number of thermal water localities occur in Presidio County; fourteen individual wells or springs with water temperatures equal to or exceeding 30°C (86°F) were located. In general, the thermal waters occur in two areas: one area is along the Rio Grande Valley, and the other is south of Marfa (FIGURE 2).

In western Presidio County, along the river valley from north to south, are the Gulf wells (locations 32-33, FIGURE 2), an unnamed spring north of Candelaria (location 31), Warm Spring (location 35), an unnamed spring east of Candelaria (location 34), and two warm springs northwest of Presidio (locations 26-27).

The two Gulf wells, north of Candelaria, were drilled to depths of 11,695 and 8,821 feet. The hot water, which reaches the surface at approximately 72°C (162°F), occurs in Cretaceous limestone at a depth of about 3000 feet (R. Kopp, personal communication, 1976). The limestones are capped by Tertiary volcanic tuffs. The warm springs south of the Gulf wells appear to rise along faults in Quaternary bolson-fill.

South of Marfa are located five thermal water occurrences, (locations 25, 28, 29, 36 and 37), with one occurrence as a spring. The waters appear to rise along faults in Tertiary volcanic units.

Brewster County

Four warm springs and two warm water wells have been found in this county. The warm springs (locations 39-42) occur on, or near, the Rio Grande floodplain. The warm waters reach the surface along faults, and are believed to originate at a depth of approximately 2000 feet in the Edwards Limestone (T. Hobbs, personal communication, 1976).

Northeast of Big Bend Park, warm water is encountered in limestone at a depth of about 900 feet; the temperature of the water is 30°C (86°F), (location 43). Water at 37°C is found at 1030 feet in a well northeast of Alpine (location 44).

Temperatures of Thermal Occurrences

The distribution of temperatures of the thermal waters in west Texas is shown in TABLE 3. Temperatures range from 30°C (86°F) to 72°C (160°F), with the hottest waters occurring in northwestern Presidio County at the Gulf wells (locations 32 and 33, FIGURE 2). The average water temperature is approximately 37°C (99°F), and the median value is 34°C (93°F). The average thermal water temperature is approximately 13°C (23°F) above the average water temperature of the region.

TABLE 4

Temperatures of Thermal Water Occurrences in Trans-Pecos, Texas

COUNTY	SAMPLES	RANGE	AVERAGE
El Paso	8	30° - 58°C	37°C
Hudspeth	10	32° - 46°C	38°C
Culberson	2	30° - 32°C	31°C
Jeff Davis	4	30° - 32°C	31°C
Presidio	14	30° - 72°C	40°C
Brewster	6	30° - 41°C	37°C

Area of Geothermal Potential

Based upon the silica geothermometer, subsurface waters above 125°C exist in seven locations in Trans-Pecos, Texas (FIGURE 3). These include areas in northeast El Paso, western and southeastern Jeff Davis, western and northern Presidio, and southern Brewster Counties. Details concerning the individual areas are given below.

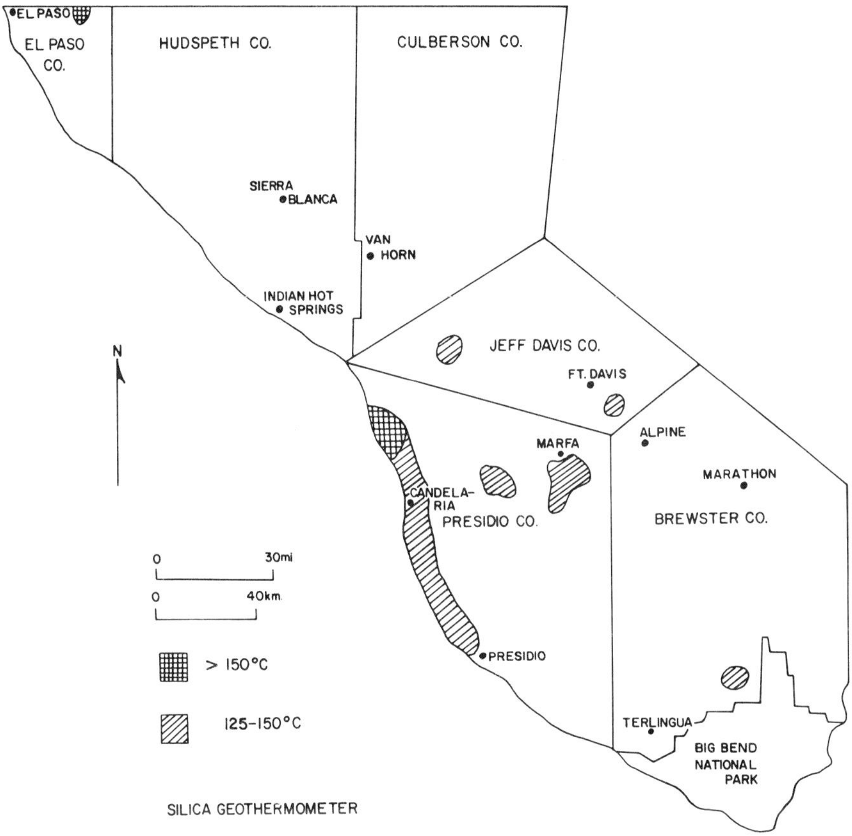

FIGURE 3. Silica Geothermometer Temperatures of Subsurface Waters, Trans-Pecos, Texas.

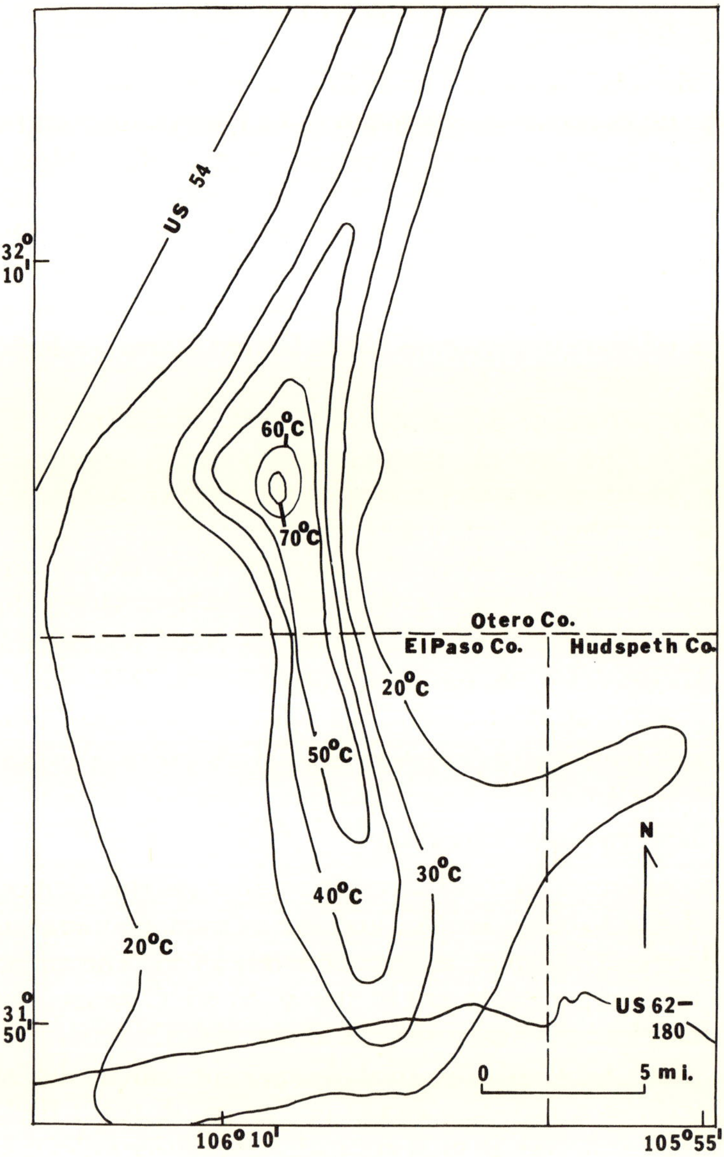

FIGURE 4. Water Temperatures, Northeast El Paso County, Texas and Southwest Otero County, New Mexico.

Northeastern El Paso County

Surface Temperatures

A thermal anomaly covering approximately ninety square miles occurs in northeastern El Paso County and extends into southwestern Otero County, New Mexico. Within this general region, 31 water wells were located, but of this total, only 21 were available for sampling. Data on the remaining ten wells were obtained from the literature (Office of Saline Waters, Report 561) or from unpublished notes of personnel at the United States Geological Survey office in El Paso.

Temperatures of well water in the area range from 18°C to 71°C. The average temperature is 28°C, nearly 5°C above the average water temperatures in west Texas.

The distribution of water temperatures is shown in FIGURE 4. A belt of thermal water (30°C or above) some three miles wide extends across the Texas-New Mexico line to a point eighteen miles north of the New Mexico State line. Maximum water temperatures of 61° to 71°C occur in New Mexico approximately five miles north of the Texas line, and temperatures as high as 58°C occur on the Texas side of the line.

Silica Geothermometer Temperatures

Temperatures derived from the silica geothermometer range from 45°C to 151°C, and the average temperature, based upon 26 samples, is approximately 78°C. The majority of the temperatures are in the 80 to 85°C range; only four samples have temperatures above 100°C.

A plan view of the distribution of silica derived temperatures is shown in FIGURE 5. Temperatures of 100°C or above occur over an area approximately three miles wide and eight miles long. The areas of highest silica temperatures correspond roughly to the areas of highest surface water temperatures.

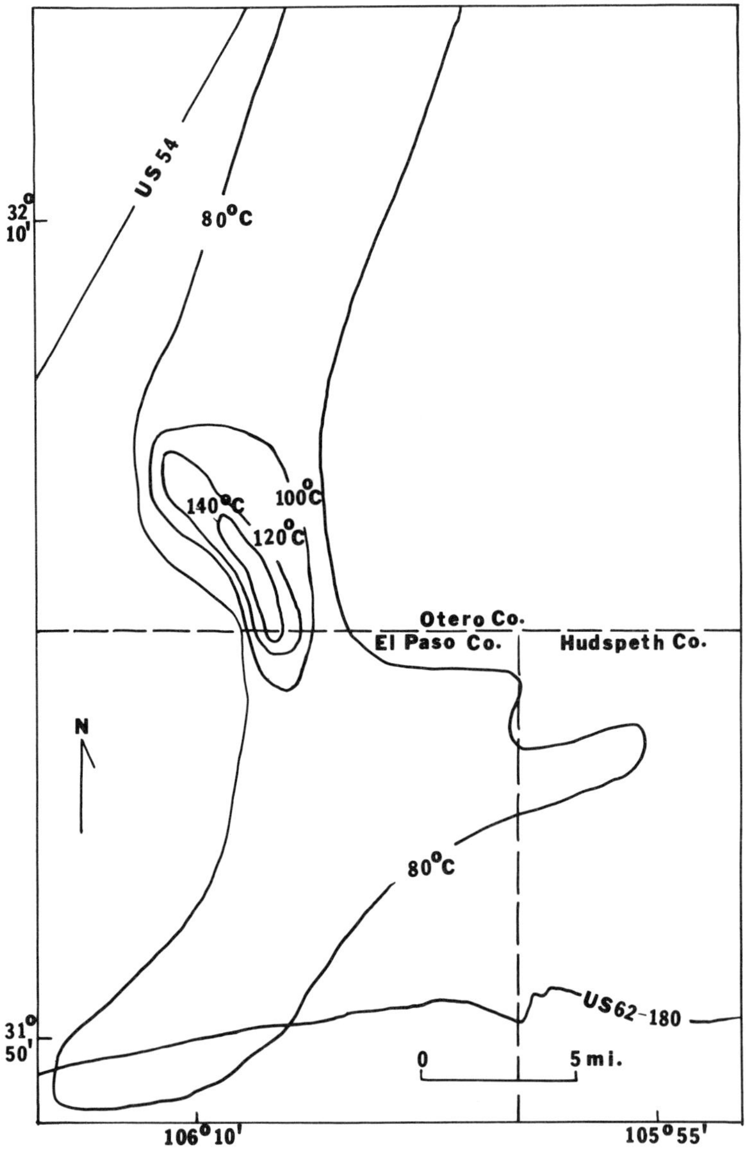

FIGURE 5. Silica Geothermometer Temperatures, Northeast El Paso County, Texas and Southwestern Otero County, New Mexico.

Thermal Gradient

The thermal gradient is a measurement of the rate increase in temperature with depth; it is a ratio of the bottom hole temperature in relation to the depth of the well. Thermal gradient values in this area, based upon calculations from fourteen wells, range from 1°C/690 ft. to a high of 1°C/7 ft. The areas of highest thermal gradient correspond to those areas of highest surface and silica temperatures.

The long, narrow appearance of all three thermal anomalies strongly suggests that hot subsurface water is reaching the surface along a north striking fault.

Western Presidio County

Surface temperatures

A belt of warm surface water, 25°C or above, extends for approximately fifty miles along the Rio Grande Valley in western Presidio County, and the width of the warm water belt averages about nine miles (FIGURE 6). Temperatures of subsurface waters range from 25 to 72°C.

Within this belt, three regions of hot water occur: 1) an area approximately fifteen miles north of Candelaria, where two oil tests have encountered subsurface water at 72 and 70°C; 2) east of Ruidoso, where a spring discharges water to the surface at 45°C, and 3) an area approximately eighteen miles southeast of Ruidoso, where two springs discharge water at 32 and 44°C.

Silica Geothermometer Temperatures

A map of the silica temperatures of this region is given in FIGURE 7. Silica temperatures range from 100°C to 177°C. A narrow belt of 125°C subsurface water extends approximately 50 miles along the western margin of the county. At the north end of the belt, the silica geothermometer indicates the presence of subsurface water temperatures of 160 to 177°C.

No information concerning thermal gradient values in the region is currently available.

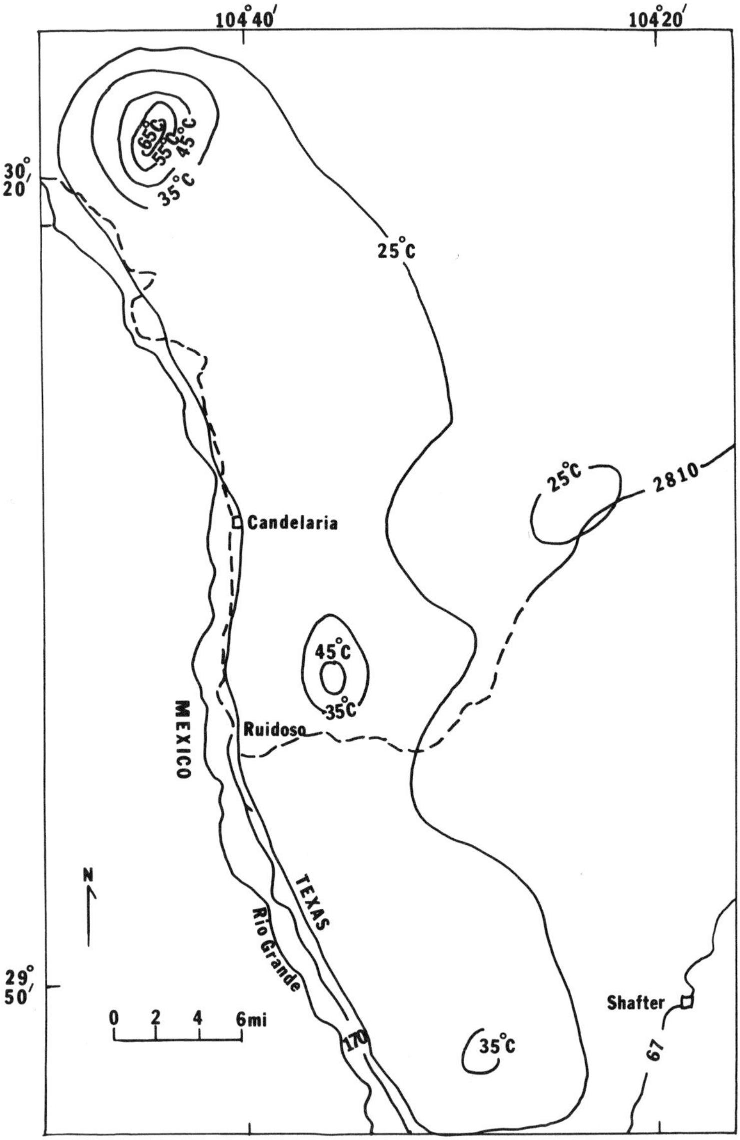

FIGURE 6. Water Temperatures, Western Presidio County, Texas.

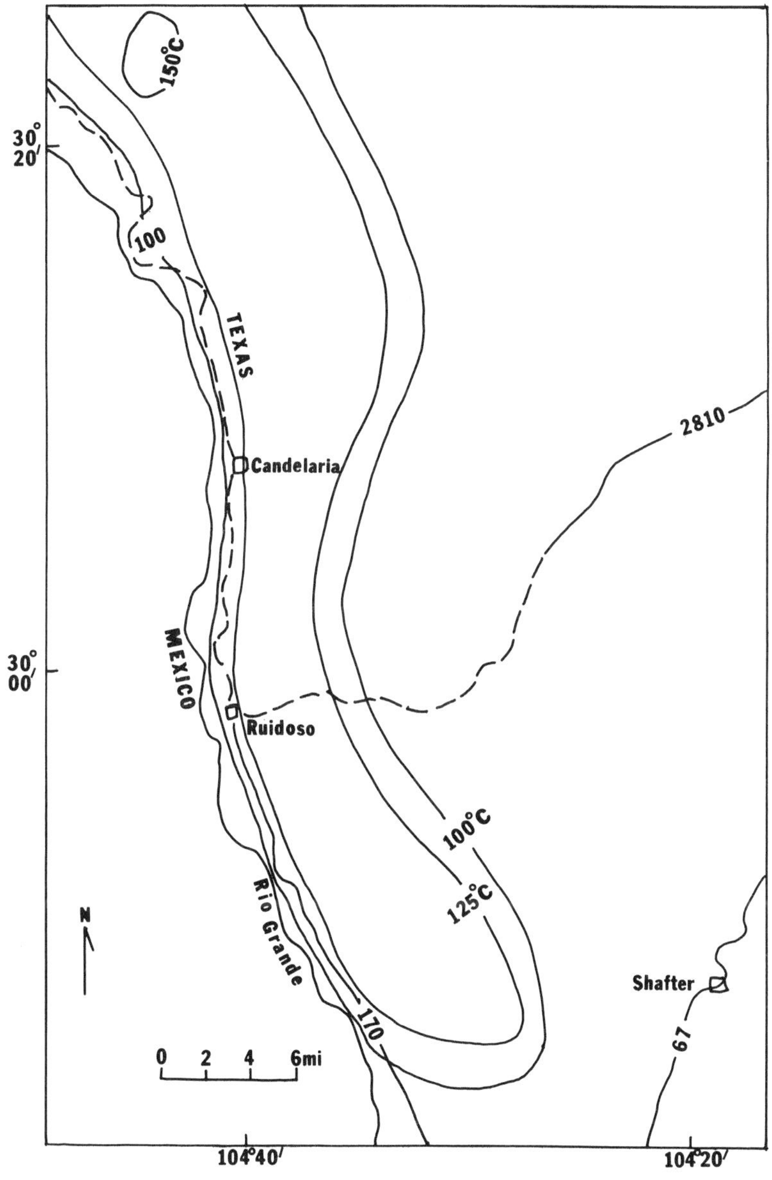

FIGURE 7. Silica Geothermometer Temperature, Western Presidio County, Texas.

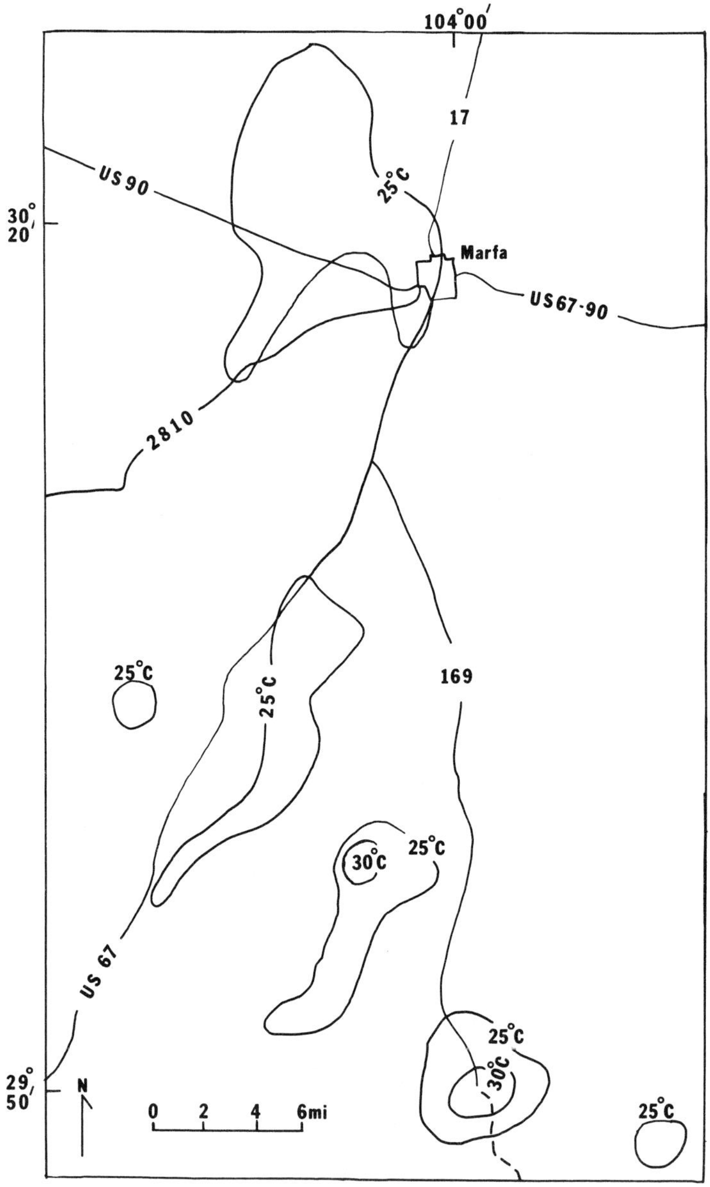

FIGURE 8. Water Temperatures, Marfa Area, Texas.

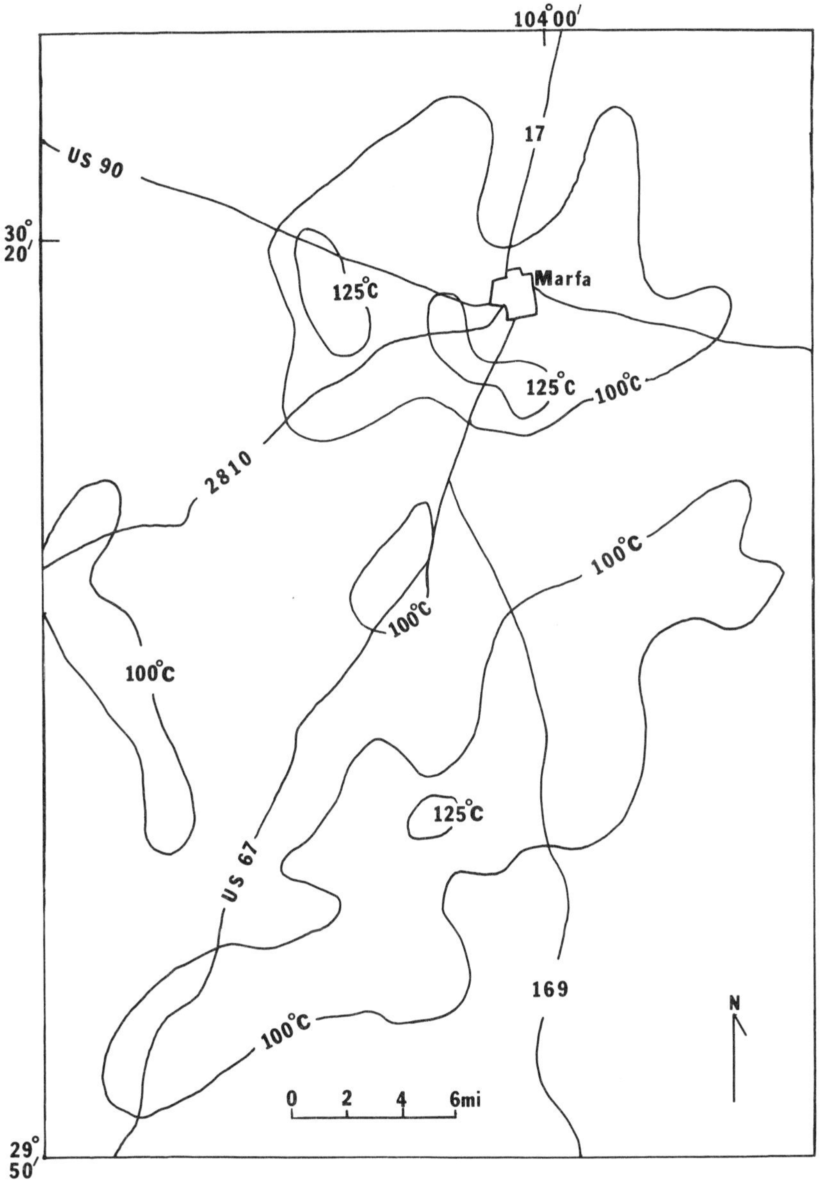

FIGURE 9. Silica Geothermometer Temperatures, Marfa Area, Texas.

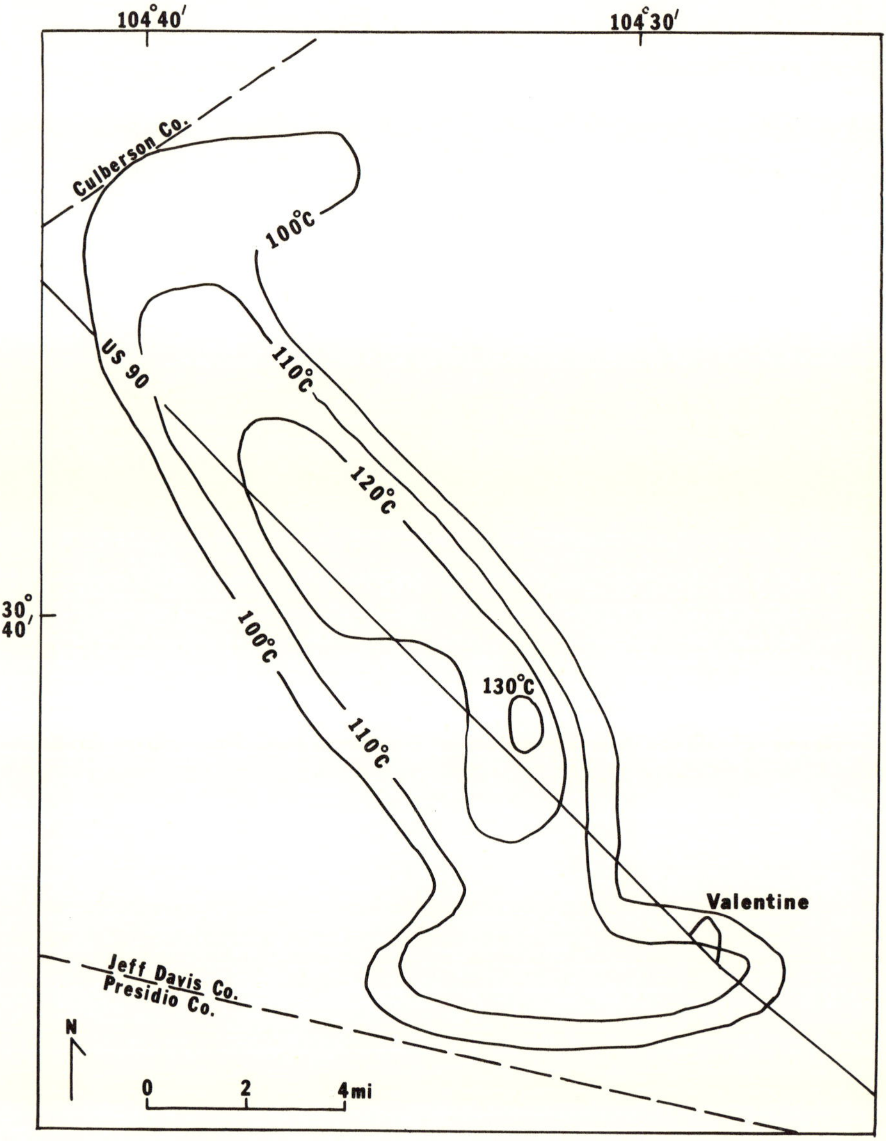

FIGURE 10. Silica Geothermometer Temperatures, Valentine Area, Texas.

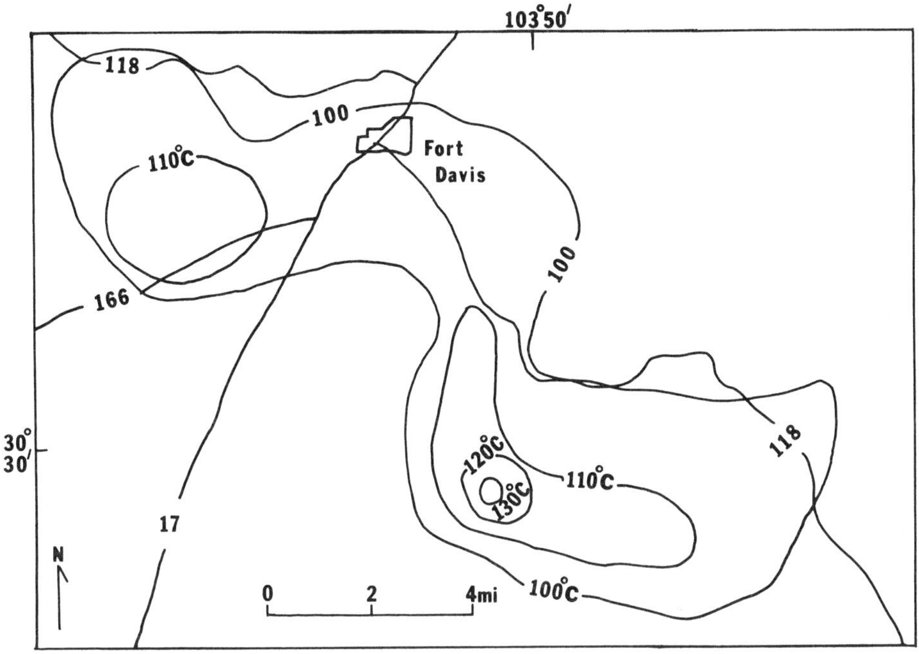

FIGURE 11. Silica Geothermometer Temperatures, Fort Davis Area, Texas.

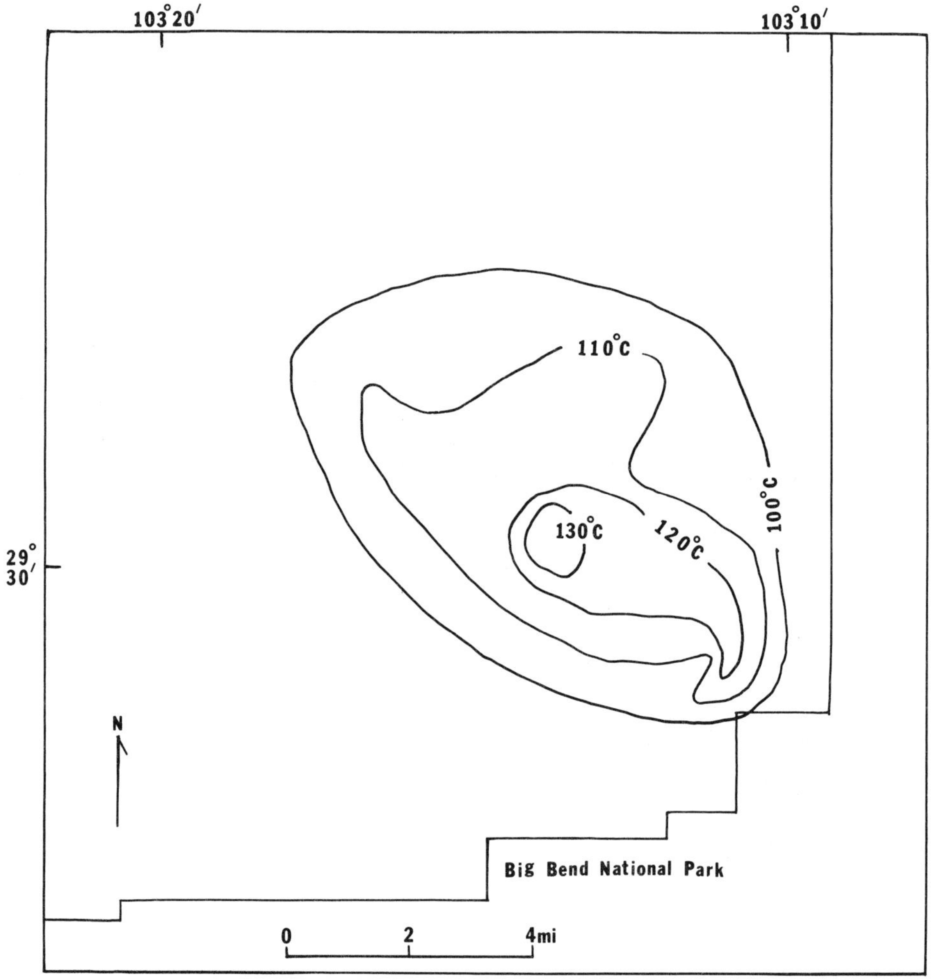

FIGURE 12. Silica Geothermometer Temperatures, Big Bend National Park Area, Texas.

Marfa Area

Surface Temperatures

Several small heat anomalies occur west and south of Marfa. Surface water temperatures equal to or greater than 25°C occur in seven minor areas, and water above 30°C occurs in three of the areas (FIGURE 8). The highest measured water temperature is 33°C.

Silica Geothermal Temperatures

Silica temperatures of 100°C or above occur in four areas, as shown in FIGURE 9. The maximum silica temperature of 130°C occurs just south and west of Marfa.

No information on thermal gradients in the area is available.

Jeff Davis and Brewster Counties

Based upon the silica geothermometer temperatures, three small areas in Jeff Davis and Brewster Counties appear to contain pockets of hot subsurface waters. The areas are 1) northwest of Valentine, Jeff Davis County (FIGURE 10), 2) south of Fort Davis, Jeff Davis County (FIGURE 11), and 3) northwest of Big Bend National Park, Brewster County (FIGURE 12). The maximum silica temperatures obtained from the three areas are 130 to 135°C. No surface temperature anomalies occur in the three areas.

Chemistry of the Thermal Waters

The thermal waters of west Texas range from fresh to slightly salty; total dissolved solids range from 275 to 11,500 ppm. On the basis of total dissolved solids, the thermal waters of the region can be divided into two groups. The first group contains the thermal waters that show greater than 5000 ppm total dissolved solids; these include well waters in eastern El Paso and western Hudspeth counties (locations 5, 6, 7, and 9) and the spring waters at Indian

Hot Springs in southern Hudspeth County, (locations 10-14). The second group contains less than 3000 ppm total dissolved solids. The major ions in the high total dissolved-solids waters are sodium and chloride, from possible solution of subsurface evaporite deposits.

In TABLE 5, the concentrations of nineteen chemical ions from all subsurface waters collected in west Texas are given, chemical analyses of the individual thermal occurrences have been reported previously (Hoffer, 1978). The following conclusions can be made concerning thermal waters of the region: 1) The average temperature of thermal waters is 37°C; this value is approximately 14°C above the average temperature of subsurface waters in west Texas. 2) Sodium, potassium, silicon, chloride, lithium, fluoride, bromide, iodide, strontium, and manganese ions are more abundant in thermal waters than in nonthermal waters. These higher concentrations are probably due to the high solubility of thermal as opposed to nonthermal waters, or selective solution of subsurface evaporite deposits by thermal waters, or both. 3) The concentration of calcium in thermal waters is significantly lower than concentration of calcium in nonthermal waters. This is probably due to decreasing solubility of calcium carbonate in solution at elevated temperature. 4) Iron, copper, zinc, and lead are less abundant in thermal waters; this is probably the result of contamination from metal pipes and casing in the more numerous sampled nonthermal wells and windmills. 5) Barium, silver, and cadmium occur in almost equal amounts in both thermal and nonthermal subsurface waters of west Texas.

TABLE 5

Temperature, pH, and Chemical Ions in Thermal Waters, Trans-Pecos, Texas

	AVERAGE (PPM)	RANGE (PPM)	WEST TEXAS AVERAGE (PPM)*
Temperature	37.1° C	30-72°C	23.2° C
pH	7.4	6.8-9.1	7.3
Sodium	501.8	6.0-2809.0	166.9
Potassium	31.9	0.0-280.0	9.3
Calcium	71.6	1.0-330.0	115.3
Silicon	23.8	9.0-151.0	18.4
Chloride	663.0	2.0-5060.0	194.7
Fluoride	2.3	0.3-7.2	1.4
Lithium	0.5	0.0-2.8	0.1
Magnesium	20.1	0.0-81.2	31.4
Bromide	1.71	0.00-18.00	1.51
Iodide	0.05	0.00-0.70	0.01
Iron	0.84	0.0-16.85	1.5
Strontium	6.2	0.0-110.0	2.9
Manganese	0.47	0.00-18.50	0.10
Lead	0.02	0.00-0.06	0.03
Zinc	0.14	0.00-1.37	0.35
Copper	0.02	0.00-0.23	0.03
Cadmium	0.00	0.00-0.01	0.00
Silver	0.00	0.00-0.05	0.00
Barium	0.11	0.00-0.77	0.12
Total Dissolved Solids	2635	275-11,500	1290

(*average of all waters in west Texas, 1028 samples).

Geochemistry of Nonthermal Groundwaters

In groundwater evaluation, the quality of water is said to be nearly as important as the quantity (Walton, 1960). The biological and chemical characteristics of groundwater determine its usefulness for domestic, agricultural, or industrial purposes.

This section presents the chemical quality of nonthermal groundwater collected from springs, wells, and windmills in Trans-Pecos, Texas.

Chemistry of Water

The chemistry of groundwater can be exceedingly complex, but the number of major dissolved constituents is quite limited (Davis and DeWiest, 1966).

The water that falls upon the surface of the earth is relatively pure, and contains only small quantities of dissolved mineral matter. As water percolates into the earth, it dissolves various minerals by contact with rock and soil. The amount and character of the mineral matter dissolved by the water will depend upon the composition of the rocks and soil, temperature, pressure, duration of contact, materials already in solution, and the chemistry of the environment (Walton, 1970).

Presentation of Analysis

The total dissolved solids contained in groundwater generally comprise only a small fraction of one percent of the total weight (Walton, 1970). Concentrations of the various dissolved minerals are expressed as ions by weight in ppm. One ppm means one part by weight of an ion in a million parts by weight of the groundwater. Concentrations may also be expressed in parts per billion, ppb.

pH is an expression of the hydrogen-ion concentration. A pH of 7.0 indicates neutral water with equal amounts of $H+$ and $OH—$ ions. Acid waters contain more abundant

H + ions, and alkaline waters a preponderance of OH-ions. Most groundwater has pH values between 5.0 (slightly acid) and 8.0 (slightly alkaline), (Walton, 1970).

The temperature of the water, measured at the surface, is expressed in degrees centigrade.

Temperature and pH

The average temperature, measured at the surface, of well or spring water in the region is 23.7°C, and the median value is 23.0°C. Measured temperatures range from 16°C to a maximum of 72°C. The highest temperatures occur in western Presidio County and in southern Brewster County, in the southwest part of Big Bend National Park.

Temperature and pH data of the waters from the region are given in TABLE 6, and a contoured map of measured surface temperatures is shown in FIGURE 13.

TABLE 6

Temperature and pH of Groundwater, Trans-Pecos, Texas

	EL PASO	CULBERSON	HUDSPETH	JEFF DAVIS	PRESIDIO	BREWSTER
Temperature:						
median	22.5°C	23.0°C	22.0°C	22.0°C	24.0°C	23.0°C
mean	21.0°C	23.1°C	21.9°C	22.6°C	24.6°C	23.5°C
range	4.0-58.0°C	17.0-34.0°C	7.0-46.0°C	16.0-32.0°C	16.0-72.0°C	17.0-41.0°C
pH:						
median	7.4	7.2	7.1	7.3	7.4	7.2
mean	7.5	7.2	7.1	7.4	7.4	7.2
range	6.1-9.1	5.9-8.9	6.6-8.2	6.6-8.5	6.0-8.9	6.4-9.1
Samples	116	166	124	161	225	236

Total Dissolved Solids

Total dissolved solids in a water sample include all solid material in solution, but it does not include suspended

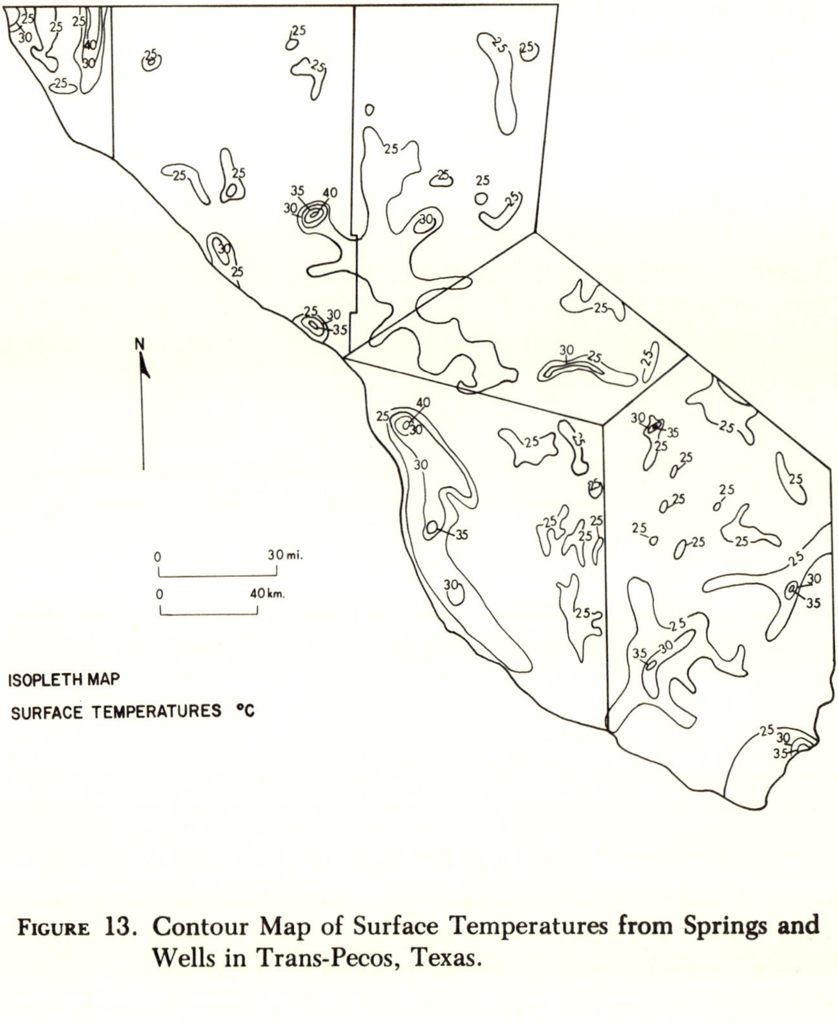

Figure 13. Contour Map of Surface Temperatures from Springs and Wells in Trans-Pecos, Texas.

materials, colloids, or dissolved gases. Natural waters range from less than 10 ppm to more than 300,000 ppm. Walton (1970) reports that range cattle in the western United States seem to be able to tolerate water with as high as 5,000 ppm of dissolved solids.

The median total dissolved solids content in the six-county area is 540 ppm and the average value is 1290 ppm. Hudspeth County groundwaters contain the highest dissolved solids content, averaging 2419 ppm, whereas the subsurface waters of Jeff Davis County average only 345 ppm. Dissolved solids data for the region are summarized in TABLE 7, and an isopleth map of total dissolved solids in groundwater of Trans-Pecos, Texas is shown in FIGURE 14.

TABLE 7

Total Dissolved Solids, ppm, Trans-Pecos, Texas Groundwaters

	EL PASO	CULBERSON	HUDSPETH	JEFF DAVIS	PRESIDIO	BREWSTER
Dissolved Solids:						
median	1350	1820	1680	255	320	490
mean	2027	2025	2419	345	521	689
range	263-11,500	200-27,000	200-11,300	80-3120	99-9000	110-9100
Samples	116	166	124	161	225	236

Major Constituents

Silicon

Silicon behind oxygen is the second most abundant element in the earth's crust. However, most minerals that contain silicon are relatively insoluble in water. Most groundwater contains from 2 to 15 ppm silicon. Water at high temperature and/or pH (i.e., pH 9.0) generally contains more abundant silicon.

The groundwaters of the six counties are moderately high in silicon, averaging 19.4 ppm, and the median value is 16 ppm. These higher-than-normal values occur because of the abundance of volcanic rocks that contain high silicon

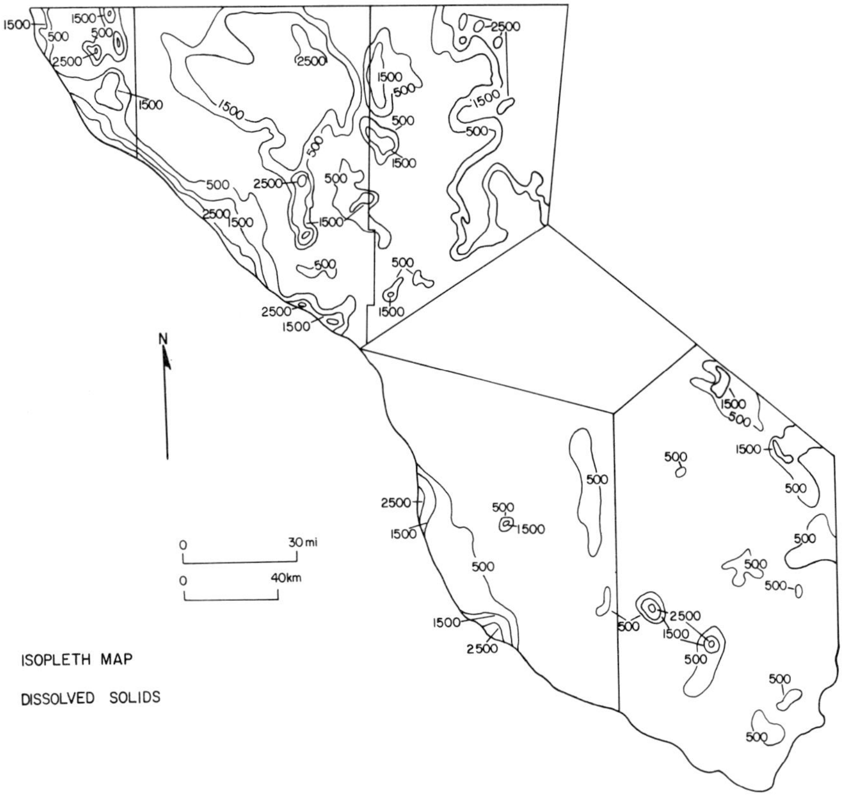

Figure 14. Isopleth Map of Total Dissolved Solids in Groundwaters of Trans-Pecos, Texas.

values. The highest silicon values are found in the warm-to-hot water springs and wells in western Presidio County.

Silicon in solution is nontoxic and interest in its presence in water is largely from operators of high pressure steam generation equipment.

TABLE 8 contains the silicon data for groundwaters of the region, and FIGURE 15 shows the distribution of silicon in groundwaters of Trans-Pecos, Texas.

TABLE 8

Silicon Concentrations, ppm, Trans-Pecos, Texas Groundwaters

	EL PASO	CULBERSON	HUDSPETH	JEFF DAVIS	PRESIDIO	BREWSTER
Silicon						
median	16	11	12	23	27	12
mean	18	12	13	23	27	15
range	9-53	1-38	3-46	5-45	1-82	1-45
Samples	116	166	124	161	225	236

Calcium

Subsurface waters derive calcium from solution of minerals such as calcite, dolomite, anhydrite, and gypsum in sedimentary rocks; and from fluorite, apatite, and feldspars in igneous and metamorphic rocks (Davis and DeWiest, 1966). Calcium, in the form of calcium carbonate, is very soluble, provided there is abundant H+; the addition of carbon dioxide increases the solubility of calcium carbonate. Concentrations of calcium in potable subsurface water generally range from 10 ppm to 100 ppm (Davis and DeWiest, 1966). Drinking water has limits of about 60 ppm (Parker, 1972). The "hardness" of water is usually expressed as the sum of its calcium and magnesium concentrations.

The average calcium concentration for groundwater in the six county area is 115 ppm, and the median value is 61 ppm. Calcium contents for subsurface waters of the in-

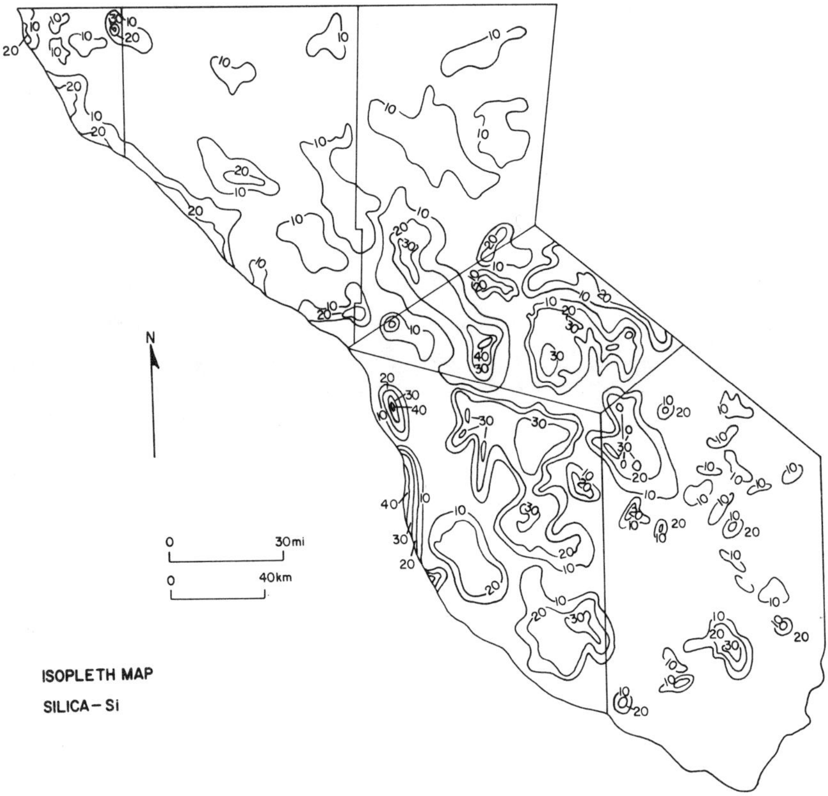

FIGURE 15. Isopleth Map of Silicon Concentrations in Groundwater, Trans-Pecos, Texas.

dividual counties are shown in TABLE 9. An isopleth map of calcium concentrations is given in FIGURE 16.

Magnesium

Magnesium is a common element that is found in all natural waters (Parker, 1972) and is the second most abundant metal in sea water. Subsurface waters contain magnesium by solution of dolomite (from sedimentary rocks); olivine, biotite, hornblende, and augite (in igneous rocks); and serpentine, talc, diopside, and tremolite (in metamorphic rocks) (Davis and DeWiest, 1966).

TABLE 9

Calcium Concentrations, ppm, Trans-Pecos, Texas Groundwaters

	EL PASO	CULBERSON	HUDSPETH	JEFF DAVIS	PRESIDIO	BREWSTER
Calcium						
mean	61.0	166.0	86.5	48.0	29.0	64.0
mean	136.5	295.7	150.9	58.6	43.0	68.0
range	1.0-1920.0	0.0-840.0	4.0-1280.0	2.0-117.0	1.0-330.0	0.0-304.0
Samples	116	166	124	147	225	236

The chemical behavior of magnesium is similar to that of calcium, but magnesium compounds are more soluble in water than are those of calcium. However, magnesium is generally found in smaller quantities in natural waters than is calcium. This is probably due to slow dissolution of dolomite and the greater abundance of calcium in the earth's crust (Davis and DeWiest, 1966).

Magnesium commonly occurs in amounts of 1 to 40 ppm in natural waters. Desirable and permissible levels of less than 60 ppm have been recommended for calcium and magnesium in drinking water (Parker, 1972).

The average magnesium concentration for groundwater in the region is 31 ppm, and the median value is 14 ppm.

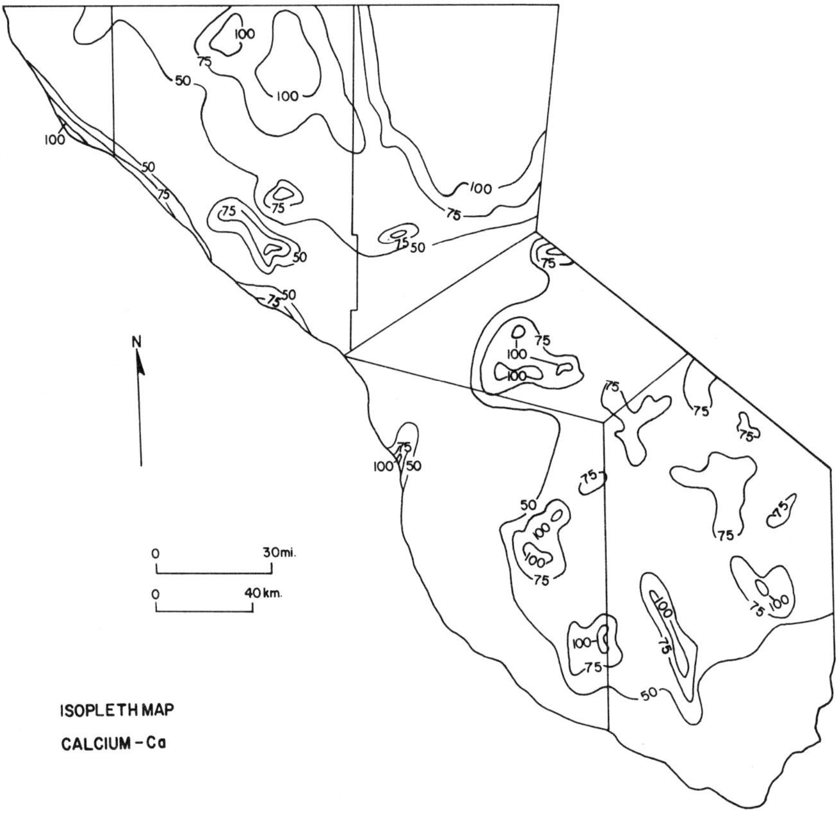

FIGURE 16. Isopleth Map of Calcium Concentrations in Groundwater, Trans-Pecos, Texas.

Magnesium values for the individual counties are shown in TABLE 10. An isopleth map of magnesium concentrations is shown in FIGURE 17.

Sodium

The primary source of sodium in natural waters is from the solution of soluble weathering products of plagioclase feldspar; locally, the solution of halite in evaporite deposits may be important (Davis and DeWiest, 1966). All natural waters contain some amount of sodium, with concentrations ranging from 0.2 pm in rain and snow to more than 100,000 ppm in brines (Davis and DeWiest, 1966).

TABLE 10

Magnesium Concentrations, ppm, Trans-Pecos, Texas Groundwaters

	EL PASO	CULBERSON	HUDSPETH	JEFF DAVIS	PRESIDIO	BREWSTER
Magnesium:						
medium	18.3	47.6	36.8	6.4	4.8	27.1
mean	24.7	68.8	53.4	8.9	6.9	35.7
range	0.1-115.9	2.4-480.0	0.1-275.6	0.0-90.4	0.0-80	0.1-1180
Samples	116	166	124	147	225	236

High levels of sodium (plus potassium) in water makes it unsuitable for the irrigation of many crops, and gives the water a salty taste when the concentration exceeds 1000 ppm.

TABLE 11

Sodium Concentrations, ppm, Trans-Pecos, Texas Groundwaters

	EL PASO	CULBERSON	HUDSPETH	JEFF DAVIS	PRESIDIO	BREWSTER
Sodium:						
median	256.0	93.0	195.5	22.0	47.0	51.0
mean	372.7	201.3	391.9	47.7	69.4	88.2
range	5.0-3490.0	1.0-6667.0	2.0-2700.0	0.0-510.0	1.0-527.0	0.0-1965.0
Samples	116	166	124	147	225	236

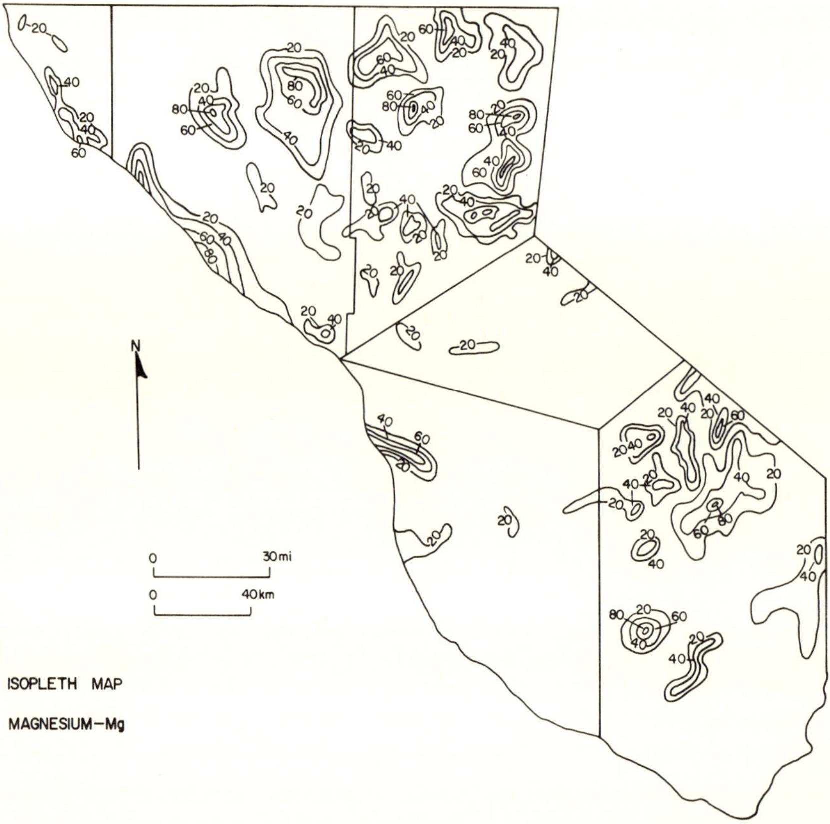

FIGURE 17. Isopleth Map of Magnesium Concentrations in Groundwater, Trans-Pecos, Texas.

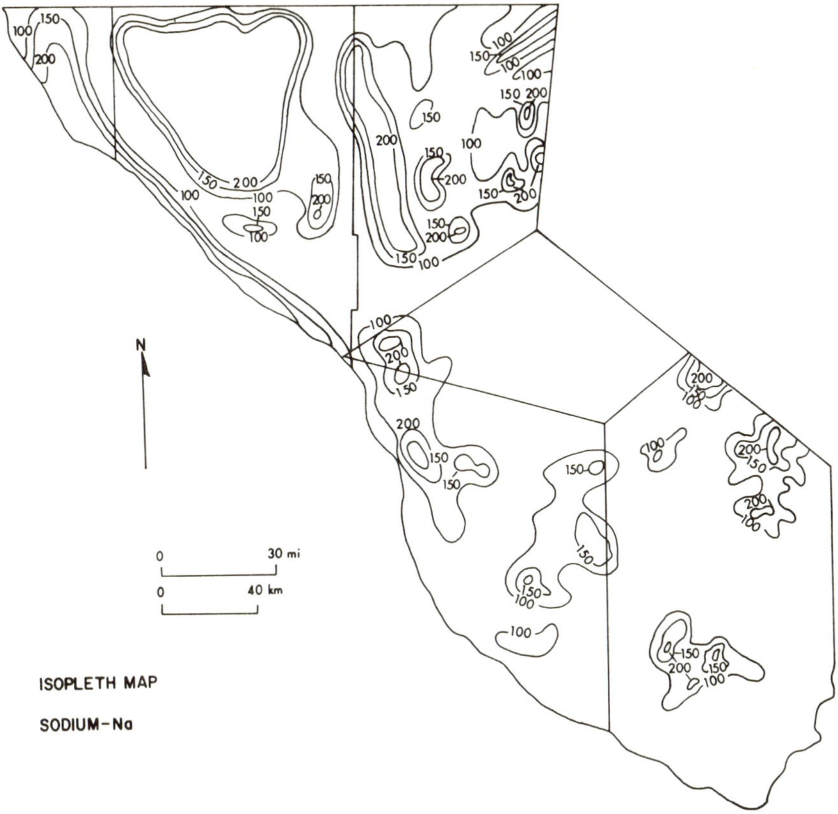

FIGURE 18. Isopleth Map of Sodium Concentrations in Groundwater, Trans-Pecos, Texas.

The average sodium content of the groundwater in the six county area is 167 ppm, and the median value is 56 ppm. Concentrations by county are given in TABLE 11; an isopleth map of sodium concentrations is given in FIGURE 18.

Chloride

Most chloride in groundwater, according to Davis and DeWiest (1966), comes from four sources: 1) chloride from ancient sea water entrapped in sediments, 2) solution of halite and related minerals in evaporite deposits, 3) concentration by evaporation of chloride contributed by rain or snow, 4) solution of dry fallout from the atmosphere. Probably the most important source is chloride transported in the atmosphere and carried to earth by precipitation.

Chloride concentrations in natural waters range from 0.1 ppm in snow to 150,000 ppm in brines; concentrations of 1000 ppm are not uncommon in subsurface waters in arid regions (Davis and DeWiest, 1966). Drinking water chloride concentration limits have been set at 250 ppm (Walton, 1972).

The average chloride concentration of groundwaters of west Texas is approximately 195 ppm, and the median value is 29 ppm. Chloride concentrations in the individual counties are given in TABLE 12; an isopleth map of chloride concentrations is shown in FIGURE 19.

TABLE 12

Chloride Concentrations, ppm, Trans-Pecos, Texas Groundwaters

	EL PASO	CULBERSON	HUDSPETH	JEFF DAVIS	PRESIDIO	BREWSTER
Chloride:						
median	257.0	64.0	104.5	13.0	19.0	23.5
mean	528.8	263.8	535.2	29.2	45.7	49.1
range	4-5060	4-15,000	2-4600	2-575	2-1000	3-990
Samples	116	166	124	147	225	236

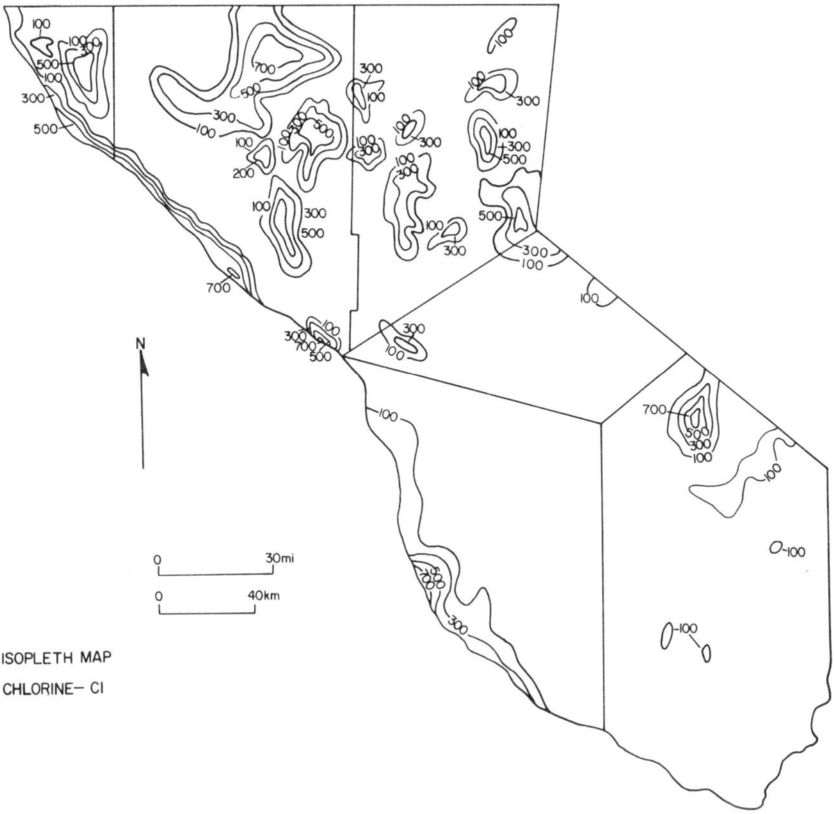

FIGURE 19. Isopleth Map of Chloride Concentrations in Groundwater, Trans-Pecos, Texas.

Secondary Constituents

Fluoride

Most of the compounds containing fluoride have low solubilities; therefore, concentrations of fluoride in natural waters are low, 0.01 to 10.0 ppm (Davis and DeWiest, 1966).

The maximum concentration level for fluoride in drinking water has been established at two times the concentration that is desirable for protecting teeth. The maximum contaminate level for fluoride is a function of the maximum daily air temperature of the location of the community water systems. These standards are given below (Hernandez and Barkley, 1976):

TEMPERATURE IN °F	LEVEL (PPM)
53.7 and below	2.4
53.8-58.3	2.2
58.4-63.8	2.0
63.8-70.6	1.8
70.6-79.2	1.6
79.2-90.5	1.4

The average fluoride concentration of groundwater in this area is 1.4 ppm, and the median value is 1 ppm. A maximum value of 19 ppm has been found in groundwaters in western Presidio county, and 20 ppm in an area southeast of Alpine in Brewster county,

Potassium

Potassium in subsurface waters is the result of partial or complete solution of potassium silicates (in igneous and metamorphic rocks) and sylvite and niter in evaporite deposits (Davis and DeWiest, 1966). The rocks of the earth's crust contain an average of approximately 25,900 ppm potassium; however, because potassium enters the structures of clay minerals, is concentrated by vegetation, and the potassium silicates are highly resistant to weathering, potassium is not very abundant in natural waters (Mason, 1960).

Most groundwater contains less than 10 ppm potassium, and commonly ranges from 1 to 5 ppm. The average potassium concentration of subsurface water in the six county area is 9.2 ppm, and the median value is 4.0 ppm. Potassium concentrations of the individual counties are given in TABLE 13.

TABLE 13

Potassium Concentrations, ppm, Trans-Pecos, Texas Groundwaters

	EL PASO	CULBERSON	HUDSPETH	JEFF DAVIS	PRESIDIO	BREWSTER
Potassium:						
median	9.0	4.0	3.5	5.0	4.0	2.0
mean	11.1	6.5	16.8	6.4	5.7	3.3
range	1-148	0-48	1-280	0-84	0-100	0-90
Samples	116	166	124	147	225	236

Iron

Iron is an abundant element in rocks of the earth's crust, averaging some 50,000 ppm (Mason, 1966). The weathering of such minerals as pyroxene, amphibole, magnetite, pyrite, biotite, and garnet produces relatively insoluble and stable iron oxides (Davis and DeWiest, 1966). Most iron in water occurs as suspended material, with very little in solution. The mean iron concentration of suspended material in surface waters in the United States is 3 ppm for dissolved iron, and the mean value is only 0.05 ppm (Kopp and Kroner, 1967).

A limit of 0.3 ppm has been set for domestic water (Parker, 1972). The limit has been set, not because iron is toxic, but to prevent brown color from forming on fixtures and laundry, and, because of the unpleasant taste in drinking water. Iron concentrations in the individual counties are given in TABLE 14.

TABLE 14

Iron Concentrations, ppm, Trans-Pecos, Texas Groundwaters

	EL PASO	CULBERSON	HUDSPETH	JEFF DAVIS	PRESIDIO	BREWSTER
Iron:						
median	0.1	0.1	0.2	0.3	0.3	0.3
mean	0.8	0.8	1.3	1.7	1.7	1.8
range	0.0-16.9	0.0-19.1	0.0-33.2	0.0-54.5	0.0-55.4	0.0-28.8
Samples	116	166	124	147	225	124

Strontium

Strontium is moderately abundant in the rocks of the earth's crust, it averages 450 ppm (Mason, 1960). It is chemically similar to calcium, and occurs in minerals in the same structural position as calcium (Davis and DeWiest, 1966). In the crust of the earth, strontium occurs most abundantly as sulphates or carbonates.

In subsurface waters, strontium concentrations range from 0.01 to 1.0 ppm, and the maximum reported value in potable waters is 52 ppm (Davis and DeWiest, 1966). A concentration of 110 ppm strontium has been measured in a warm water well in northern Brewster County. In Presidio County, south of Marfa, concentrations of strontium up to 46 ppm have been measured in waters of several warm springs.

Strontium concentrations in subsurface waters of the individual counties are listed in TABLE 15.

TABLE 15

Strontium Concentrations, ppm, Trans-Pecos, Texas Groundwaters

	EL PASO	CULBERSON	HUDSPETH	JEFF DAVIS	PRESIDIO	BREWSTER
Strontium:						
medium	1.32	3.80	2.26	0.4	0.4	0.8
mean	3.55	6.97	4.00	1.5	1.4	1.6
range	0.01-54.89	0.19-79.70	0.01-36.66	0.0-54.7	0.01-76.00	0.0-110.0
Samples	116	166	124	147	225	236

Minor Constituents

Barium

Barium occurs as sulphates and carbonates in crustal rocks; its abundance in the crust is approximately 400 ppm. An arbitrary limit of 1 ppm has been set for drinking water because of the toxic effect of barium on the heart, nervous system, and blood vessels (Parker, 1972).

Concentrations of barium in the subsurface waters of Trans-Pecos, Texas average 0.12 ppm. Barium concentrations in the individual counties are shown in TABLE 16.

TABLE 16

Barium Concentrations, ppm, Trans-Pecos, Texas Groundwaters

	EL PASO	CULBERSON	HUDSPETH	JEFF DAVIS	PRESIDIO	BREWSTER
Barium:						
median	0.09	0.09	0.09	0.10	0.09	0.09
mean	0.07	0.18	0.08	0.10	0.11	0.11
range	0.00-0.33	0.00-18.10	0.00-0.55	0.00-0.33	0.00-0.77	0.00-0.33
Samples	116	166	124	147	225	236

Lithium

Lithium is relatively rare in the earth's crust, averaging only 30 ppm (Mason, 1960). It occurs most abundantly in micas in granitic rocks. Concentrations of lithium in groundwater range from 0.001 to 0.5 ppm, but brines may exceed 5 ppm (Davis and DeWiest, 1966).

The average lithium content in subsurface waters of the region is 0.06 ppm, and the highest measured value is 3.10 ppm.

Lithium concentrations in the subsurface waters of the area are given in TABLE 17.

TABLE 17

Lithium Concentrations, ppm, Trans-Pecos, Texas Groundwaters

	EL PASO	CULBERSON	HUDSPETH	JEFF DAVIS	PRESIDIO	BREWSTER
Lithium:						
median	0.11	0.06	0.10	0.00	0.00	0.00
mean	0.17	0.16	0.27	0.05	0.07	0.06
range	0.00-1.34	0.00-1.95	0.00-3.10	0.00-0.5	0.00-0.8	0.00-1.6
Samples	116	166	124	147	225	236

Bromide and Iodide

Bromide and iodide are ions of halogen elements; they average 3 ppm and 0.3 ppm in crustal rocks (Mason, 1966). Bromide is fairly abundant in sea water, comprising approximately 0.2 percent of the dissolved constituents. Iodide is concentrated by plants and animals, so that sea water is very low in that element.

Concentrations of bromide and iodide in subsurface waters of the region average 1.5 and 0.9 ppm, and the median values are 0.4 for bromide and 0.0 ppm for iodide.

Copper, Lead, and Zinc

Copper, lead, and zinc average approximately 45, 15, and 65 ppm in rocks of the earth's crust. They occur predominantly with sulfur as sulphide minerals. These elements are not generally reported in routine analyses, but they have been measured by several workers as an aid in prospecting for mineral deposits (Davis and DeWiest, 1966). High temperature acid waters commonly contain the largest concentrations of these elements.

TABLE 18 lists the criteria for the various uses of water in terms of the tolerable amounts of copper, zinc, and lead.

TABLE 18

Recommended Limits, in ppm, for Copper, Lead and Zinc in Water

ELEMENT	DRINKING WATER	SHORT TERM USE	CONTINUOUS USE
Copper	1.0	5.0	0.2
Lead	0.05	20.0	5.0
Zinc	5.0	10.0	5.0

The average concentrations for these elements in subsurface waters of the region are 0.03 ppm for copper, 0.03 ppm for lead, and 0.35 ppm for zinc.

Manganese

The chemical behavior of manganese is very similar to that of iron, but its occurrence in natural waters is generally less than half the value of iron (Davis and DeWiest, 1966). Rocks of the earth's crust average about 1000 ppm manganese (Mason, 1960).

As with iron, objections to manganese arise because of the coloration imparted to the water, not because of toxicity. The following recommended concentrations have been suggested: 1) drinking water - 0.05 ppm, 2) irrigation waters (short term use - 20 ppm; continuous use - 2 ppm) (Parker, 1972).

Manganese concentrations for groundwaters of each county are given in TABLE 19.

TABLE 19

Manganese Concentrations, ppm, Trans-Pecos, Texas Groundwaters

	EL PASO	CULBERSON	HUDSPETH	JEFF DAVIS	PRESIDIO	BREWSTER
Manganese:						
median	0.03	0.01	0.01	0.0	0.0	0.0
mean	0.28	0.06	0.16	0.2	0.1	0.04
range	0.0-4.55	0.0-1.51	0.0-4.80	0.0-0.65	0.0-18.5	0.0-2.64
Samples	116	166	124	147	225	236

Cadmium

Cadmium is found only occasionally, in trace quantities, in natural waters. Kopp and Krosner (1967) report the presence of cadmium in less than 3% of the samples analyzed by the U. S. Geological Survey.

The maximum permissible level for drinking water is 0.01 ppb and the recommended levels for irrigation waters are between 0.05 and 0.005 ppb (Parker, 1972).

Cadmium concentrations for groundwaters of the individual counties are shown in TABLE 20.

TABLE 20

Cadmium Concentrations, ppm, Trans-Pecos, Texas Groundwaters

	EL PASO	CULBERSON	HUDSPETH	JEFF DAVIS	PRESIDIO	BREWSTER
Cadmium:						
median	0.00	0.00	0.00	0.00	0.00	0.00
mean	0.00	0.00	0.00	0.00	0.00	0.00
range	0.00-0.03	0.00-0.02	0.00-0.22	0.00-0.01	0.00-0.01	0.00-0.01
Samples	116	166	124	147	225	236

Trace Constituents

Silver

Silver is of only minor abundance in rocks of the earth's crust, averaging 0.1 ppm (Mason, 1960). In natural waters, silver is found in concentrations between 0.0001 and .001 ppm; the average content is about 0.0002 ppm (Parker, 1972).

TABLE 21 summarizes the silver concentrations of the individual amounts.

TABLE 21

Silver Concentrations, ppm, Trans-Pecos, Texas Groundwaters

	EL PASO	CULBERSON	HUDSPETH	JEFF DAVIS	PRESIDIO	BREWSTER
Silver:						
median	0.00	0.00	0.00	0.00	0.00	0.00
mean	0.00	0.01	0.00	0.00	0.00	0.00
range	0.00-0.09	0.00-0.07	0.00-0.05	0.00-0.02	0.00-0.1	0.00-0.01
Samples	116	166	124	147	225	236

Future Studies

Programs are in progress at The University of Texas at El Paso to define more precisely the boundaries and extent of the hot water reservoirs in Trans-Pecos, Texas. In addition, studies are under way to relate the chemical constituents in groundwater, as presented here, to the geology of the region.

ACKNOWLEDGEMENTS

The author extends his thanks to Dr. Arleigh B. Templeton, President, and Dr. Kenneth E. Beasley, Vice-President, The University of Texas at El Paso, for obtaining funds to complete the study.

Field samples were collected by Tom Hobbs, John Potter, Ken Carraway, and Richard Kopp, all geology graduate students, U. T. El Paso. Mr. Robert Curtis, also a graduate student at U. T. El Paso, did most of the chemical analyses.

REFERENCES

Davis, S. N. and R. J. M. DeWiest.
1966. *Hydrogeology*. John Wiley and Sons, Inc., 461 pp.

Fournier, R. O. and J. J. Rowe.
1966. "Estimation of Underground Temperatures from the Silica Content of Water from Hot Springs and Wet-steam Wells," *American Journal of Science*, 264, pp. 685-97.

Groat, C. G..
1972. "Presidio Bolson, Trans-Pecos, Texas and Adjacent Mexico: Geology of a Desert Basin Aquifer System," University of Texas at Austin, Bureau of Economic Geology, Report of Investigation 76, 46 pp.

Hernandez, J. W. and W. A. Barkley (eds.).
1976. "Proceedings of the Twenty-first Annual New Mexico Water Conference," *Water Resources Research Institute Report No. 72*, 125 pp.

Hoffer, J. M.
1978. "Thermal Water Occurrences in Trans-Pecos, Texas," *Texas Journal of Science*, *XXX*, pp. 309-319

Jones, B. R.
1968. "Geology of Southern Quitman Mountains and Vicinity, Hudspeth County, Texas," Texas A & M University, Ph.D dissertation, 162 pp.

Kopp, J. F. and R. C. Droner.
1967. "Trace Metals in Waters of the United States," U. S. Department of Interior, F.W.P.C.A., Division of Pollution Surveillance.

Kruger, P. and C. Otte (eds.).
1973. *Geothermal Energy*, Stanford University Press, 360 pp.

Mason, B.
1960. *Principles of Geochemistry*, John Wiley and Sons, 310 pp.

Parker, C. R.
1972. *Water Analysis by Atomic Absorption Spectroscopy*, Varian Techtron, 78 pp.

Reaser, D. F., J. R. Underwood and B. R. Jones.
1975. "Geothermal Prospects of the Eagle-Quitman Mountains Vicinity Trans-Pecos, Texas," *West Texas Geological Society Guidebook*, pp. 155-61.

Summers, W. K.
1976. "Catalog of Thermal Waters in New Mexico," *New Mexico Bureau of Mines and Mineral Resources, Hydrologic Report No. 4*, 80 pp.

Walton, W. C.
1970. *Groundwater Resource Evaluation*. McGraw-Hill, Inc., 664 pp.

Warning, G. A.
1965. "Thermal Springs of the United States and Other Countries of the World, a Summary," *U. S. Geological Survey Professional Paper 492*, 383 pp.

EDITORIAL ASSISTANT:
David Timmons

COVER DESIGN:
Kathleen Rogers

TYPOGRAPHY:
Lenore Locke

PRODUCTION:
UTEP Printing Division